We hope this book has been informative and helpful on your journey to understanding and celebrating older adults. Thank you for your interest and support!

Title: Evolution of Hydrogen Power: The Future of Clean Energy Transportation

Subtitle: Fuel Cells and Zero Emissions

Series: Ride Through Time: The Story of World Vehicles

By M.J. Knightly

"The transportation industry is on the brink of a revolution, with advancements in autonomous vehicles, electric cars, and new modes of transportation like hyperloops and vertical takeoff and landing aircrafts."
Mary Barra, CEO of General Motors

"Innovation in transportation is not only about speed and efficiency, but also about sustainability and reducing our impact on the environment."
Patricia Espinosa, Executive Secretary of the United Nations Framework Convention on Climate Change

"The future of transportation will be about creating a seamless and integrated experience, where different modes of transportation work together to get people where they need to go."
Dara Khosrowshahi, CEO of Uber

"The age of electric vehicles is here, and it's exciting to see how technology is driving the transformation of our transportation systems."
Elon Musk, CEO of Tesla and SpaceX

"Transportation is not just about getting from point A to point B, it's about the freedom and opportunities that come with mobility. Innovations in transportation are critical for empowering individuals and communities around the world."
Ban Ki-moon, former Secretary-General of the United Nations

"As transportation continues to evolve, we have the opportunity to create a more equitable and accessible system that benefits all people, regardless of their background or circumstances."
Keisha Lance Bottoms, former mayor of Atlanta, Georgia

"Innovation is key to unlocking the potential of transportation, and we need to encourage and invest in new ideas and technologies that will shape the future of mobility."
Hiroto Saikawa, former CEO of Nissan Motor Company

Table of Contents

Introduction

The importance of clean energy in transportation

The transportation sector is a major contributor to global greenhouse gas emissions, and as such, it has become a focal point for efforts to mitigate climate change. The use of traditional fossil fuels in transportation has been identified as a major source of carbon emissions, which contributes significantly to the adverse effects of climate change such as global warming, sea level rise, and extreme weather conditions. Therefore, the shift towards clean energy in transportation has become a pressing need. In this chapter, we will discuss the importance of clean energy in transportation and how it can mitigate climate change.

The Importance of Clean Energy in Transportation:

Clean energy is defined as energy that comes from renewable or low-carbon sources, and it has become a key solution to reduce greenhouse gas emissions and combat climate change. Transportation is responsible for around 14% of global greenhouse gas emissions, with most of it coming from the burning of fossil fuels in vehicles. This makes the transportation sector a key area for the deployment of clean energy technologies.

The adoption of clean energy in transportation can provide several benefits, including:

1. Reduction in greenhouse gas emissions: By switching to clean energy sources, we can significantly reduce the amount of carbon emissions from the transportation sector. This can help us mitigate climate change and avoid its catastrophic effects.

2. Improved air quality: Traditional fossil fuel-powered vehicles produce harmful pollutants such as nitrogen oxides, particulate matter, and sulfur dioxide. These pollutants have adverse health effects on humans and contribute to smog and acid rain. The use of clean energy in transportation can reduce these pollutants, leading to better air quality and public health.

3. Energy security: The use of clean energy in transportation can enhance energy security by reducing dependence on fossil fuels, which are often imported from unstable regions of the world. By using local and renewable energy sources, we can increase energy independence and reduce the geopolitical risks associated with fossil fuel imports.

4. Job creation: The shift towards clean energy in transportation can create new job opportunities in industries such as renewable energy production, vehicle manufacturing, and infrastructure development.

In conclusion, clean energy in transportation is essential for mitigating climate change, improving air quality, enhancing energy security, and creating new job opportunities. The transportation sector plays a vital role in the transition towards a low-carbon economy, and the adoption of clean energy technologies is a key solution to reduce greenhouse gas emissions.

The rise of hydrogen fuel cells as a potential solution

Hydrogen fuel cells have emerged as a promising solution for clean energy transportation, offering several advantages over traditional fossil fuel-powered engines. In this chapter, we will discuss the rise of hydrogen fuel cells as a potential solution for clean energy transportation and their advantages over traditional engines.

Hydrogen fuel cells are electrochemical devices that convert the chemical energy of hydrogen and oxygen into electrical energy, producing water as the only byproduct. They are a promising technology for clean energy transportation, as they offer several advantages over traditional fossil fuel-powered engines.

1. Zero emissions: Hydrogen fuel cells emit only water vapor, making them a zero-emission alternative to traditional engines. This can significantly reduce greenhouse gas emissions and improve air quality.

2. High efficiency: Hydrogen fuel cells have a higher energy conversion efficiency than traditional engines, making them a more efficient energy source.

3. Versatility: Hydrogen can be produced from a variety of sources, including renewable sources such as wind and solar, making it a versatile energy source.

4. Fast refueling: Hydrogen fuel cells can be refueled in a matter of minutes, similar to traditional gasoline-powered vehicles, unlike battery electric vehicles that require much longer charging times.

5. Low noise: Hydrogen fuel cells operate quietly, providing a quieter driving experience than traditional engines.

The potential of hydrogen fuel cells for clean energy transportation has been recognized by governments and industry players worldwide, leading to significant investment in research and development.

Despite these advantages, there are also some challenges associated with the adoption of hydrogen fuel cells in transportation. One of the main challenges is the lack of hydrogen infrastructure, including refueling stations and production facilities. Additionally, the high cost of hydrogen fuel cells and their components remains a barrier to widespread adoption.

In conclusion, hydrogen fuel cells have emerged as a promising solution for clean energy transportation, offering several advantages over traditional fossil fuel-powered engines. While there are still some challenges that need to be addressed, the potential of hydrogen fuel cells for clean energy transportation is significant, and continued investment and development can help overcome these challenges and realize their full potential.

The goal of the book to explore the promise and challenges of hydrogen power

The goal of this book is to explore the promise and challenges of hydrogen power as a clean energy source for transportation. Hydrogen fuel cells have the potential to revolutionize the transportation sector, offering a zero-emission alternative to traditional fossil fuel-powered engines. However, there are also significant challenges associated with the adoption of hydrogen fuel cells, including the lack of infrastructure and high cost.

The Promise of Hydrogen Power:

Hydrogen fuel cells offer several benefits that make them a promising solution for clean energy transportation. As mentioned earlier, they emit only water vapor, making them a zero-emission alternative to traditional engines. This can significantly reduce greenhouse gas emissions and improve air quality. Additionally, hydrogen is a versatile energy source that can be produced from a variety of sources, including renewable sources such as wind and solar.

Another benefit of hydrogen fuel cells is their high energy conversion efficiency, which makes them a more efficient energy source than traditional engines. This can result in cost savings over the long term, despite the initial high cost of fuel cells and their components.

Furthermore, hydrogen fuel cells offer fast refueling times, similar to traditional gasoline-powered vehicles, which can help overcome the range anxiety associated with battery electric vehicles. Hydrogen fuel cells also operate quietly, providing a quieter driving experience than traditional engines.

The Challenges of Hydrogen Power:

Despite the promise of hydrogen power, there are significant challenges that need to be addressed for its widespread adoption. One of the main challenges is the lack of infrastructure, including refueling stations and production facilities. This is a significant barrier to adoption, as drivers need access to refueling stations in order to use hydrogen-powered vehicles. Additionally, the cost of hydrogen fuel cells and their components remains high, which limits their adoption.

There are also challenges associated with hydrogen production and storage. Hydrogen is highly flammable and requires specialized storage and handling procedures, which can be expensive and complicated.

Another challenge is the potential environmental impact of widespread adoption. While hydrogen fuel cells emit only water vapor, the production of hydrogen itself can generate greenhouse gas emissions if not produced from renewable sources.

The Goal of the Book:

The goal of this book is to provide a comprehensive exploration of the promise and challenges of hydrogen power as a clean energy source for transportation. By examining the current state of research and development in hydrogen-powered vehicles and infrastructure, as well as the potential for future advancements, we can gain a better understanding of the potential of hydrogen power for clean energy transportation. Additionally, by exploring the challenges and opportunities associated with hydrogen power, we can identify potential solutions and strategies to address these challenges and advance the adoption of hydrogen fuel cells in transportation.

In conclusion, the goal of this book is to explore the promise and challenges of hydrogen power as a clean energy source for transportation. While there are significant challenges associated with the adoption of hydrogen fuel cells, their potential for zero-emission transportation is significant, and continued investment and development can help overcome these challenges and realize their full potential.

Chapter 1: Understanding Hydrogen Fuel Cells
What are hydrogen fuel cells and how do they work?

Hydrogen fuel cells are an innovative technology that can generate electricity by combining hydrogen with oxygen, producing only water and heat as byproducts. Fuel cells are based on the principle of electrochemical reaction, where the conversion of chemical energy into electrical energy takes place directly.

A fuel cell consists of several layers, including the anode, cathode, and electrolyte membrane. The anode is where hydrogen is introduced into the fuel cell, and the cathode is where oxygen enters. The electrolyte membrane, which is a thin, permeable layer, separates the anode and cathode and allows protons to move from the anode to the cathode.

As hydrogen molecules enter the fuel cell, they react with a catalyst, which splits them into protons and electrons. The protons move through the electrolyte membrane to the cathode, while the electrons travel through an external circuit, creating an electric current. At the cathode, oxygen molecules combine with the protons and electrons to form water and release heat energy.

Hydrogen fuel cells are highly efficient, with typical efficiency rates of 40-60%. This is significantly higher than the 25-35% efficiency of traditional gasoline engines. Additionally, fuel cells are quiet, produce no greenhouse gas emissions or other pollutants, and are highly reliable with few moving parts.

There are different types of fuel cells, including proton exchange membrane (PEM), solid oxide fuel cell (SOFC), alkaline fuel cell (AFC), and molten carbonate fuel cell (MCFC). Each type has its advantages and disadvantages, and each has different applications. PEM fuel cells are the most common type used in transportation due to their high power density and low operating temperatures. SOFCs are better suited for stationary power generation because of their high efficiency and ability to use a variety of fuels.

Overall, hydrogen fuel cells have the potential to transform the energy landscape by providing a clean, efficient, and reliable source of power. While there are still challenges to overcome in terms of cost, infrastructure, and technology development, the promise of hydrogen power is significant, and ongoing research and development efforts aim to make it a viable solution for the future.

The advantages and disadvantages of hydrogen as a fuel source

Hydrogen is a versatile and abundant element that has been touted as a potential solution for many of the world's energy and environmental challenges. As a fuel source, hydrogen offers several advantages over traditional fossil fuels, as well as some notable disadvantages. In this section, we will explore the benefits and drawbacks of hydrogen as a fuel source.

Advantages of Hydrogen Fuel Cells

1. Clean Energy: One of the most significant advantages of hydrogen fuel cells is that they produce only water and heat as byproducts, making them a clean energy source. Hydrogen fuel cells are free of pollutants such as carbon dioxide, nitrogen oxide, and sulfur dioxide, which are known to cause environmental damage and contribute to climate change.

2. High Efficiency: Hydrogen fuel cells are highly efficient, with an energy conversion efficiency of up to 60%. This is significantly higher than the conversion efficiency of traditional gasoline engines, which typically have an efficiency of around 20%.

3. Abundant Resource: Hydrogen is the most abundant element in the universe, making it a virtually

inexhaustible resource. Additionally, hydrogen can be produced from a variety of sources, including water, biomass, and fossil fuels.

4. Versatility: Hydrogen can be used in a variety of applications, including transportation, power generation, and heating. Hydrogen fuel cells can power anything from small electronic devices to large vehicles and buildings.

Disadvantages of Hydrogen Fuel Cells

1. High Cost: One of the biggest challenges facing hydrogen fuel cell technology is its cost. Hydrogen fuel cell vehicles and infrastructure are still relatively expensive, and the cost of producing and storing hydrogen is also high.

2. Safety Concerns: Hydrogen is a highly flammable gas, which raises safety concerns in the production, storage, and transportation of hydrogen.

3. Lack of Infrastructure: Hydrogen fuel cell vehicles require a specialized refueling infrastructure that is not yet widely available. Building a network of hydrogen refueling stations is a significant challenge that will require significant investment and time.

4. Energy Intensive Production: Producing hydrogen is an energy-intensive process that requires large amounts of electricity. If this electricity is generated from fossil fuels, it

can undermine the environmental benefits of hydrogen fuel cells.

Conclusion

Despite the challenges facing hydrogen fuel cell technology, it offers several significant advantages as a clean and versatile energy source. As research and development continue, the cost of hydrogen fuel cells is likely to decrease, and the infrastructure necessary to support them will become more widely available. With continued innovation and investment, hydrogen fuel cells have the potential to play a critical role in the transition to a more sustainable and environmentally friendly energy system.

The current state of hydrogen fuel cell technology

Hydrogen fuel cell technology has been advancing rapidly in recent years, with significant investments being made by both the public and private sectors. The technology has been applied in a variety of settings, including transportation, stationary power, and portable devices. In this section, we will explore the current state of hydrogen fuel cell technology, including its strengths and weaknesses, as well as the current barriers to its widespread adoption.

Advancements in Hydrogen Fuel Cell Technology

One of the main areas of focus in the development of hydrogen fuel cell technology has been to improve its efficiency and reliability. Over the years, there have been significant advancements in the materials used in fuel cells, which has led to greater efficiency in converting hydrogen into electricity. In addition, there have been significant improvements in the durability and longevity of fuel cells, making them more practical for use in a variety of applications.

One of the most notable advancements in recent years has been the development of high-temperature fuel cells, which operate at temperatures above 800°C. These fuel cells use ceramic materials as the electrolyte, which allows for greater efficiency and lower costs. High-temperature fuel cells also have the potential to be used in a wider range of applications, including distributed power generation and fuel cell vehicles.

Another area of advancement has been in the development of fuel cell stacks, which are made up of multiple fuel cells that are connected together. These stacks have been designed to be more compact and efficient, which has led to greater power density and lower costs.

Current Limitations and Challenges

Despite these advancements, there are still several challenges that need to be addressed before hydrogen fuel cells can be widely adopted as a clean energy source. One of the main challenges is the cost of producing hydrogen, which is still relatively high compared to traditional fossil fuels. This is due in part to the current lack of infrastructure for hydrogen production and distribution, which limits the economies of scale that could drive down costs.

Another challenge is the need for greater durability and reliability in fuel cells. While significant advancements have been made in this area, fuel cells still have a limited lifespan, and their performance can degrade over time. This can result in increased maintenance costs and a shorter operating life, which can make them less attractive compared to other clean energy technologies.

Finally, there is also a need for greater public awareness and acceptance of hydrogen fuel cell technology. Many people are still unfamiliar with the technology and its potential benefits, and there is a need for greater education and outreach to help build public support.

Conclusion

The current state of hydrogen fuel cell technology is promising, with significant advancements being made in efficiency, durability, and reliability. However, there are still several challenges that need to be addressed before the technology can be widely adopted as a clean energy source. The development of more efficient and cost-effective hydrogen production and distribution infrastructure, as well as greater public awareness and support, will be critical in the continued advancement and adoption of this technology.

The potential of hydrogen as a renewable energy source

Hydrogen is considered a renewable energy source because it can be produced from renewable sources of energy such as wind, solar, and hydropower. When produced through electrolysis, hydrogen can be considered a zero-emission fuel, since the process only produces water and oxygen as byproducts. This makes hydrogen a promising solution to combat climate change and reduce greenhouse gas emissions.

One of the main benefits of hydrogen as a renewable energy source is that it is abundant. Hydrogen is the most abundant element in the universe and can be found in virtually unlimited quantities on Earth. While it is typically found in compounds such as water and hydrocarbons, it can be extracted and isolated through various methods, including steam methane reforming, electrolysis, and biomass gasification.

Another advantage of hydrogen as a renewable energy source is its versatility. Hydrogen can be used in fuel cells to generate electricity, in combustion engines to power vehicles, and in industrial processes to produce chemicals such as ammonia and methanol. Its flexibility means that it has the potential to replace fossil fuels in a wide range of applications, from transportation to heating and power generation.

Hydrogen also has the advantage of being a clean fuel, emitting no greenhouse gases or pollutants when burned in fuel cells or combustion engines. This makes it an attractive alternative to traditional fossil fuels, which are major contributors to air pollution and climate change.

However, there are also challenges associated with the use of hydrogen as a renewable energy source. One major challenge is the cost of production and storage. While the cost of producing hydrogen has been declining in recent years, it is still more expensive than fossil fuels in many cases. Additionally, hydrogen is difficult to store and transport due to its low density, which requires large, specialized tanks or pipelines.

Another challenge is the need for infrastructure to support the use of hydrogen. This includes refueling stations for hydrogen-powered vehicles, pipelines for transporting hydrogen, and storage facilities for large-scale use in power generation and industrial applications. Building this infrastructure requires significant investment and coordination between governments, private industry, and other stakeholders.

Despite these challenges, the potential of hydrogen as a renewable energy source is significant. With continued research and development, hydrogen could play a key role in the transition to a cleaner, more sustainable energy future.

Chapter 2: The History of Hydrogen in Transportation

The early history of hydrogen as a fuel source

The use of hydrogen as a fuel source dates back to the early 19th century, when it was first discovered by English scientist Henry Cavendish. However, it wasn't until the late 1800s that scientists began exploring the potential of hydrogen as a fuel for transportation.

In 1807, Francois Isaac de Rivaz, a Swiss inventor, developed the first internal combustion engine, which ran on a mixture of hydrogen and oxygen. The engine was designed to power a vehicle, but it was too heavy and inefficient to be practical. Nonetheless, de Rivaz's invention laid the groundwork for future development of hydrogen-powered vehicles.

In the late 1800s and early 1900s, hydrogen was used as a fuel for airships, which were used for transportation and military purposes. The German airship Hindenburg, which famously caught fire and crashed in 1937, was powered by hydrogen.

In the mid-20th century, interest in hydrogen as a fuel source for transportation began to grow again. During World War II, the German military developed hydrogen fuel cell technology as a power source for submarines. In the 1960s,

NASA began using hydrogen fuel cells to power spacecraft, including the Apollo missions.

In the 1970s and 1980s, a number of automakers began experimenting with hydrogen-powered vehicles. General Motors developed a prototype hydrogen fuel cell vehicle in 1966, and in the 1980s, Ford and Chrysler also developed hydrogen-powered concept cars. However, these vehicles were never commercialized, largely due to the high cost of producing hydrogen fuel cells at the time.

In recent years, the development of more efficient and cost-effective fuel cell technology has renewed interest in hydrogen as a fuel source for transportation. Today, a number of automakers, including Toyota, Hyundai, and Honda, offer hydrogen fuel cell vehicles for sale to consumers. Additionally, companies such as Nikola and Tesla are developing hydrogen-powered trucks and semi-trucks for commercial use.

Overall, the early history of hydrogen as a fuel source for transportation laid the foundation for the development of modern hydrogen fuel cell technology. While early experiments with hydrogen-powered vehicles were limited by the high cost and inefficiency of the technology, recent advancements in fuel cell technology have made hydrogen a more viable and promising fuel source for transportation.

The role of hydrogen in space exploration

Hydrogen has played a critical role in space exploration since the 1950s. In fact, hydrogen fuel cells were first developed for the U.S. space program as a power source for spacecraft. Today, hydrogen remains an essential part of space exploration and the development of new technologies for space travel.

One of the earliest uses of hydrogen in space exploration was in the launch vehicles for the Apollo moon missions. The Saturn V rocket, which was used to launch the Apollo spacecraft, was powered by liquid hydrogen and liquid oxygen. This fuel combination was chosen because it has a high specific impulse, which means it provides a lot of thrust for a given amount of fuel. This was critical for launching the heavy Apollo spacecraft and getting it into orbit around the Earth and on its way to the moon.

Hydrogen fuel cells were also used on the Apollo missions to provide electrical power to the spacecraft. The fuel cells combined hydrogen and oxygen to generate electricity, with water as the only byproduct. This was a critical technology for the Apollo missions, as it provided a reliable source of power for the spacecraft during their journeys to the moon.

Hydrogen has continued to play a key role in space exploration in the decades since the Apollo missions. For example, the Space Shuttle, which was in use from 1981 to 2011, used hydrogen fuel cells to generate electricity for the spacecraft's systems. The International Space Station (ISS) also uses hydrogen fuel cells as a backup power source, in addition to solar panels.

Beyond spacecraft power systems, hydrogen has also been used in other aspects of space exploration. For example, hydrogen has been used as a propellant for rocket engines, including the engines used on the Saturn V rocket and the Space Shuttle. Hydrogen has also been proposed as a fuel source for future space missions, including missions to Mars and beyond.

Overall, the role of hydrogen in space exploration has been critical to the development of new technologies and the advancement of our understanding of the universe. The use of hydrogen as a fuel source and power generation technology has proven to be reliable and effective, and will continue to be an important part of future space missions.

The development of hydrogen fuel cell vehicles

Hydrogen fuel cell vehicles have been in development for decades, with significant progress made in recent years. The development of these vehicles has been driven by concerns about the environmental impact of traditional fossil fuel-powered vehicles, as well as the need to reduce dependence on foreign oil. In this section, we will explore the history and progress of hydrogen fuel cell vehicles.

Early Development of Hydrogen Fuel Cell Vehicles The concept of using hydrogen as a fuel source dates back to the early 1800s, but it wasn't until the 1960s that serious research and development of fuel cells began. NASA was one of the first organizations to use fuel cells, as they were ideal for providing power on spacecraft due to their high energy density and low weight.

In the 1970s and 1980s, automakers began to experiment with using fuel cells to power vehicles. However, progress was slow due to the high cost of fuel cells and the lack of infrastructure for refueling. The first hydrogen fuel cell vehicle was developed in 1993 by General Motors, but it was not until the early 2000s that significant progress was made in the development of hydrogen fuel cell vehicles.

Recent Progress in Hydrogen Fuel Cell Vehicles In recent years, there has been significant progress in the development of hydrogen fuel cell vehicles. Major automakers such as Toyota, Honda, and Hyundai have released hydrogen fuel cell vehicles for consumer use, with more models expected in the near future.

One of the key advantages of hydrogen fuel cell vehicles is their range. Unlike battery-electric vehicles, which require frequent charging, hydrogen fuel cell vehicles can travel hundreds of miles on a single tank of hydrogen. Additionally, hydrogen fuel cell vehicles emit only water vapor, making them a truly zero-emission transportation option.

However, there are still challenges to be addressed in the development of hydrogen fuel cell vehicles. One of the main challenges is the cost of fuel cells, which remain prohibitively expensive for many consumers. Additionally, there is still a lack of hydrogen refueling infrastructure, which limits the practicality of hydrogen fuel cell vehicles for many consumers.

Future Outlook for Hydrogen Fuel Cell Vehicles Despite these challenges, there is reason to be optimistic about the future of hydrogen fuel cell vehicles. The potential for these vehicles to reduce greenhouse gas emissions and dependence on fossil fuels is significant, and advances in technology and infrastructure are likely to make them more practical and affordable in the coming years.

Governments and private industry are investing heavily in the development of hydrogen fuel cell vehicles and infrastructure. For example, California has set a goal of having 200 hydrogen refueling stations by 2025, and automakers are working to reduce the cost of fuel cells and increase their durability and reliability.

In conclusion, the development of hydrogen fuel cell vehicles has been a long and challenging process, but significant progress has been made in recent years. While there are still challenges to be addressed, the potential of hydrogen fuel cell vehicles to provide a zero-emission transportation option is significant, and the future looks bright for this promising technology.

The current landscape of hydrogen-powered transportation

As interest in clean energy grows, more attention is being paid to the potential of hydrogen as a fuel source for transportation. In recent years, several companies and governments around the world have been working to develop hydrogen-powered vehicles and build the necessary infrastructure to support them.

Current State of Hydrogen-Powered Vehicles

The current landscape of hydrogen-powered transportation is still in its early stages, with only a small number of hydrogen fuel cell vehicles (FCVs) available for purchase or lease. These vehicles are primarily passenger cars, but there are also some buses and commercial trucks powered by hydrogen fuel cells.

At the forefront of the FCV market are the Toyota Mirai, Honda Clarity, and Hyundai Nexo. These vehicles use hydrogen fuel cells to convert hydrogen and oxygen into electricity to power an electric motor, emitting only water vapor as a byproduct.

One of the biggest advantages of FCVs is their long driving range, which is comparable to that of gasoline-powered vehicles. For example, the Toyota Mirai has a range of around 312 miles, while the Hyundai Nexo has a range of around 380 miles.

However, the high cost of producing FCVs and the lack of hydrogen refueling infrastructure have been major obstacles to their widespread adoption. As of 2021, there were only a few hundred hydrogen refueling stations worldwide, mostly located in Japan, California, and Europe.

Government Initiatives

To promote the adoption of FCVs, many governments have implemented policies to support their development and deployment. For example, in Japan, the government has set a goal of having 800,000 FCVs on the road by 2030, and has provided subsidies to help reduce the cost of purchasing FCVs.

In California, the government has also implemented a range of incentives and regulations to promote the adoption of FCVs. The state has set a goal of having 200 hydrogen refueling stations by 2025, and has provided grants to support the construction of new stations.

In Europe, several countries have implemented policies to promote the adoption of FCVs, including tax incentives, grants for purchasing FCVs, and funding for research and development.

Challenges and Opportunities

Despite the challenges facing the widespread adoption of FCVs, there are many opportunities for growth in the hydrogen-powered transportation sector. One of the most promising areas is the development of hydrogen fuel cell buses and trucks, which have the potential to greatly reduce emissions from public transportation and logistics.

Another opportunity is the development of green hydrogen, which is produced using renewable energy sources such as wind or solar power. Green hydrogen has the potential to significantly reduce emissions from hydrogen production, making it a more sustainable fuel source.

Conclusion

While the current landscape of hydrogen-powered transportation is still in its early stages, there is a growing interest in the potential of hydrogen as a fuel source. With the support of governments and continued investment in research and development, the future of hydrogen-powered transportation looks promising. However, there are still many challenges that need to be addressed, including the high cost of producing FCVs and the lack of hydrogen refueling infrastructure.

Chapter 3: The Pros and Cons of Hydrogen Fuel Cell Vehicles

The benefits of hydrogen fuel cell vehicles over traditional gasoline vehicles

Hydrogen fuel cell vehicles have a number of benefits over traditional gasoline vehicles. Here are some of the most significant:

1. Zero Emissions: One of the most obvious benefits of hydrogen fuel cell vehicles is that they produce zero emissions. Unlike gasoline vehicles, which emit harmful pollutants into the air, hydrogen fuel cell vehicles only emit water vapor and heat. This means that they can play a significant role in reducing air pollution and combating climate change.

2. Fuel Efficiency: Hydrogen fuel cell vehicles are more fuel efficient than gasoline vehicles. This is because they convert up to 60% of the energy stored in hydrogen into motion, while gasoline vehicles only convert about 20% of the energy stored in gasoline into motion. This means that hydrogen fuel cell vehicles can go farther on a single tank of fuel, making them more practical for long distance travel.

3. Quiet Operation: Hydrogen fuel cell vehicles are much quieter than traditional gasoline vehicles. This is because they do not have an internal combustion engine,

which produces noise and vibrations. This makes for a more pleasant and peaceful driving experience, especially in urban areas.

4. Fast Refueling: Hydrogen fuel cell vehicles can be refueled in just a few minutes, similar to traditional gasoline vehicles. This is much faster than electric vehicles, which can take hours to recharge. This means that hydrogen fuel cell vehicles can be more practical for drivers who need to make long trips or who don't have access to charging infrastructure.

5. Versatility: Hydrogen fuel cell vehicles can be used for a wide range of applications, from cars and trucks to buses and trains. This means that they have the potential to replace a large portion of the vehicles currently on the road, reducing emissions and improving air quality.

Despite these benefits, there are also some disadvantages to hydrogen fuel cell vehicles that must be considered:

1. Cost: Hydrogen fuel cell vehicles are currently more expensive than traditional gasoline vehicles. This is because the technology is still in the early stages of development and production volumes are low. As production volumes increase and the technology becomes more mature, costs are expected to come down.

2. Infrastructure: Hydrogen fuel cell vehicles require a network of hydrogen refueling stations to be practical for everyday use. Currently, there are only a few dozen hydrogen refueling stations in the United States, making it difficult for drivers to find a place to refuel. However, this infrastructure is expected to grow as more hydrogen fuel cell vehicles are produced.

3. Energy Efficiency: Although hydrogen fuel cell vehicles are more fuel efficient than gasoline vehicles, they are not as energy efficient as battery electric vehicles. This is because the process of producing hydrogen is not as energy efficient as simply storing electricity in a battery. This means that hydrogen fuel cell vehicles may not be the best choice for all applications, especially where energy efficiency is a top priority.

4. Safety: Hydrogen is a highly flammable gas, which raises some safety concerns. However, the tanks used to store hydrogen in fuel cell vehicles are designed to be extremely safe, with multiple layers of protection to prevent leaks and explosions.

Overall, the benefits of hydrogen fuel cell vehicles are significant and show promise for a more sustainable future. However, there are also some challenges that need to be addressed to make this technology more practical and widely adopted.

The potential environmental impact of widespread adoption

The potential environmental impact of widespread adoption of hydrogen fuel cell vehicles is a topic of significant interest and debate. On one hand, hydrogen fuel cells are often touted as a zero-emissions solution for transportation, which could significantly reduce greenhouse gas emissions and air pollution. On the other hand, the production, storage, and transportation of hydrogen can have its own environmental impacts.

One of the primary benefits of hydrogen fuel cell vehicles over traditional gasoline vehicles is their potential to significantly reduce greenhouse gas emissions. When hydrogen is produced using renewable energy sources such as wind or solar power, it can be considered a completely emissions-free fuel source. Even when hydrogen is produced using fossil fuels, such as natural gas, the emissions associated with hydrogen production are generally lower than those associated with traditional gasoline production.

In addition to reducing greenhouse gas emissions, hydrogen fuel cell vehicles have the potential to reduce other air pollutants as well. For example, nitrogen oxides (NOx) and particulate matter (PM) are significant contributors to air pollution in urban areas, and hydrogen fuel cell vehicles emit little to no NOx or PM. This could have significant health benefits, particularly for individuals living in urban areas with high levels of air pollution.

However, the potential environmental benefits of hydrogen fuel cell vehicles are not without their challenges. One significant challenge is the production of hydrogen. While hydrogen can be produced using renewable energy sources, most hydrogen today is produced using fossil fuels such as natural gas. The production of hydrogen from fossil fuels produces carbon dioxide (CO_2) emissions, which can offset some of the environmental benefits of hydrogen fuel cell vehicles.

Additionally, the transportation and storage of hydrogen can also have environmental impacts. Hydrogen is typically stored and transported at high pressures, which can result in leaks and other safety concerns. The pipelines used to transport hydrogen can also have their own environmental impacts, particularly if they are located in sensitive ecosystems or areas with high levels of community opposition.

Another potential environmental impact of widespread adoption of hydrogen fuel cell vehicles is the impact on water resources. Hydrogen fuel cell vehicles produce water vapor as their primary emissions, which is generally considered a benign byproduct. However, the production of hydrogen requires significant amounts of water, particularly if it is produced using electrolysis. In areas with limited water resources, the production of hydrogen could exacerbate existing water scarcity concerns.

In conclusion, the potential environmental impact of widespread adoption of hydrogen fuel cell vehicles is complex and multifaceted. While these vehicles have the potential to significantly reduce greenhouse gas emissions and air pollution, the production, storage, and transportation of hydrogen can have their own environmental impacts. It is important to carefully consider these impacts when evaluating the potential role of hydrogen fuel cell vehicles in our transportation system.

The challenges of infrastructure development and maintenance

One of the main challenges of widespread adoption of hydrogen fuel cell vehicles (FCVs) is the development and maintenance of a robust hydrogen infrastructure. Unlike gasoline or diesel fuel, hydrogen is not yet widely available at fueling stations. This lack of infrastructure has been a major barrier to the deployment of FCVs in many parts of the world. In this section, we will explore the challenges associated with infrastructure development and maintenance for hydrogen fuel cell vehicles.

1. Building an Infrastructure for Hydrogen Fueling Stations

One of the main challenges in developing a hydrogen fueling infrastructure is the cost of building hydrogen fueling stations. Hydrogen fueling stations require specialized equipment and materials, which can be expensive to manufacture and install. The cost of building a single hydrogen fueling station can range from $1 to $2 million dollars, depending on the location and design.

2. The Role of Government in Infrastructure Development

To overcome the challenges associated with building a hydrogen fueling infrastructure, many governments around the world have started to invest in infrastructure development. In the United States, the Department of Energy has set a goal of deploying at least 100 hydrogen fueling stations by 2025. In Japan, the government has launched a plan to build 900 hydrogen fueling stations by 2030. These investments are expected to help drive down the cost of building and operating hydrogen fueling stations over time.

3. Maintenance and Safety of Hydrogen Fueling Stations

In addition to the initial cost of building hydrogen fueling stations, there are also ongoing costs associated with maintenance and safety. Hydrogen is a highly flammable gas, which means that fueling stations must be designed and maintained to strict safety standards. This can increase the cost of operating a hydrogen fueling station, and also requires specialized training for station operators and maintenance personnel.

4. The Challenge of Scaling Up Infrastructure

Another challenge associated with infrastructure development for hydrogen fuel cell vehicles is the need to scale up infrastructure to meet demand. As more FCVs are deployed on the roads, there will be a greater need for hydrogen fueling stations. This will require a significant investment in infrastructure, and will also require coordination between vehicle manufacturers, fuel providers, and government agencies.

5. The Role of Alternative Fuels

Finally, it is worth noting that hydrogen fuel cell vehicles are not the only option for reducing emissions from the transportation sector. Alternative fuels such as electric vehicles and biofuels also have the potential to significantly reduce emissions. In some cases, these alternative fuels may be a more practical and cost-effective solution than hydrogen fuel cell vehicles. As such, it is important to consider the full range of options when evaluating the potential of hydrogen as a fuel source for transportation.

Conclusion:

In conclusion, the development and maintenance of a hydrogen fueling infrastructure is one of the biggest challenges associated with widespread adoption of hydrogen fuel cell vehicles. The high cost of building and maintaining fueling stations, combined with the need to scale up infrastructure to meet demand, make this a difficult challenge to overcome. However, with continued government investment and collaboration between vehicle manufacturers and fuel providers, it is possible to develop a robust and sustainable hydrogen fueling infrastructure that can support the widespread adoption of hydrogen fuel cell vehicles in the future.

The cost-effectiveness of hydrogen fuel cell technology

Hydrogen fuel cell technology is considered a promising alternative to conventional fossil fuels due to its potential for zero-emissions and improved energy efficiency. However, the technology is not yet commercially competitive with traditional gasoline vehicles due to its higher cost.

The cost-effectiveness of hydrogen fuel cell technology depends on several factors, including the cost of production, storage, transportation, and infrastructure development. At present, the high cost of production is the most significant factor limiting the widespread adoption of hydrogen fuel cell vehicles.

One of the main challenges facing the industry is the cost of producing hydrogen. Currently, most hydrogen is produced through a process called steam methane reforming (SMR), which involves converting natural gas into hydrogen through a high-temperature reaction. While this process is efficient, it is also expensive, and it produces greenhouse gases as a byproduct.

Another approach to hydrogen production is through renewable energy sources such as wind and solar power, which are emissions-free. However, the technology to produce hydrogen from renewable sources is still in its early stages, and it is currently more expensive than SMR.

The cost of storing and transporting hydrogen is another significant factor in the cost-effectiveness of fuel cell vehicles. Hydrogen is a lightweight gas and has a low energy density, meaning that it must be compressed or liquefied to be transported and stored effectively. Compression and liquefaction require energy and add to the overall cost of the fuel.

The development of a comprehensive infrastructure for hydrogen fuel cell vehicles is also a challenge. Currently, there are only a few dozen hydrogen refueling stations worldwide, making it difficult for fuel cell vehicle owners to travel long distances or even across certain regions. Building out a comprehensive infrastructure of refueling stations and hydrogen production facilities will require significant investment, and this investment will likely be recouped over the long-term.

Finally, the cost-effectiveness of hydrogen fuel cell technology is heavily influenced by economies of scale. As more vehicles and infrastructure are developed, the cost of production, storage, and transportation is expected to decrease. Therefore, the future cost-effectiveness of hydrogen fuel cell technology will depend heavily on the industry's ability to attract investment and scale up production.

In summary, the cost-effectiveness of hydrogen fuel cell technology is a significant factor in its widespread adoption as a renewable energy source. While the technology has the potential to be cost-competitive with traditional gasoline vehicles over the long-term, the current high cost of production and infrastructure development poses a significant challenge. The industry is making progress in reducing costs, but significant investment is still needed to develop a comprehensive hydrogen infrastructure and drive economies of scale.

Chapter 4: The State of Hydrogen Infrastructure
The current state of hydrogen refueling stations

As hydrogen fuel cell vehicles continue to gain popularity, the availability of hydrogen refueling stations is becoming increasingly important. In this chapter, we will explore the current state of hydrogen refueling stations, their distribution, and the challenges that the industry faces in scaling up the infrastructure.

At the end of 2020, there were about 90 publicly accessible hydrogen refueling stations in the United States, according to the U.S. Department of Energy (DOE). California leads the way with over 40 operational stations, followed by Hawaii, Connecticut, and Massachusetts. There are also a few dozen stations in Europe and Asia, primarily in Japan and South Korea.

While this number may seem small compared to the over 168,000 gasoline stations in the U.S. alone, it is important to note that the growth of hydrogen refueling stations has been significant in recent years. In 2010, there were only nine hydrogen refueling stations worldwide, according to the International Energy Agency (IEA). This number had grown to 153 by 2019, and it continues to increase.

One of the biggest challenges facing the development of hydrogen refueling infrastructure is the high cost of building and maintaining the stations. According to a report by the National Renewable Energy Laboratory (NREL), the cost of building a hydrogen refueling station can range from $1 million to $5 million, depending on factors such as location, capacity, and technology used. The cost of hydrogen production and delivery is also a factor, as hydrogen must be transported and stored at high pressure, requiring specialized equipment and infrastructure.

Another challenge is the lack of standardized refueling protocols. There are currently three different types of refueling protocols in use: SAE J2601, ISO 14687, and CSA B.52. Each protocol has different specifications for the pressure, temperature, and flow rate of hydrogen, which can lead to confusion for consumers and make it difficult to develop a cohesive infrastructure.

To address these challenges, industry and government organizations are working together to develop solutions. For example, the DOE's H2USA initiative is working to accelerate the development of hydrogen refueling infrastructure and promote the adoption of fuel cell vehicles. The Fuel Cell and Hydrogen Energy Association (FCHEA) is also working to promote the development of a standardized refueling protocol.

In addition, there are a number of private companies working to build out the hydrogen refueling infrastructure. For example, Air Liquide, a global industrial gas supplier, has partnered with Toyota to build 12 hydrogen refueling stations in the Northeastern United States. And Shell is working to develop hydrogen refueling stations in Europe and the United States, with a goal of having 10 hydrogen refueling stations operational by the end of 2021.

In conclusion, while the current state of hydrogen refueling infrastructure is still in its early stages, there has been significant progress in recent years. The industry faces challenges in scaling up the infrastructure due to the high cost of building and maintaining stations, as well as the lack of standardized refueling protocols. However, with the collaboration of industry and government organizations, as well as private companies, the future of hydrogen refueling infrastructure looks promising.

The challenges of expanding infrastructure

As the adoption of hydrogen fuel cell vehicles increases, so does the need for hydrogen refueling infrastructure. However, the development and expansion of such infrastructure present several challenges that must be addressed to ensure the growth of the industry.

1. High Costs: One of the biggest challenges facing the development of hydrogen infrastructure is the high cost associated with building refueling stations. Compared to gasoline stations, hydrogen stations are significantly more expensive to build and operate. The high costs are due to the specialized equipment required to handle hydrogen, including compressors, storage tanks, dispensers, and safety systems.

2. Lack of Public Funding: Another challenge is the lack of public funding for hydrogen infrastructure. Most of the early hydrogen stations were built with government grants, and private investors have been hesitant to invest in infrastructure due to high costs and uncertainty regarding the long-term viability of the technology.

3. Siting and Permitting: Siting and permitting can also be challenging for hydrogen refueling stations. Local zoning and building codes can vary significantly from one

jurisdiction to another, which can make it difficult to build stations in a timely and cost-effective manner.

4. Limited Public Awareness: There is also a limited public awareness of hydrogen fuel cell technology and its benefits. This lack of awareness can make it difficult to gain support for the expansion of infrastructure and can hinder the adoption of hydrogen fuel cell vehicles.

5. Lack of Standardization: There is also a lack of standardization in the design and operation of hydrogen refueling stations. This lack of standardization can make it difficult for manufacturers to produce equipment that is compatible with all stations, which can hinder the growth of the industry.

6. Safety Concerns: The storage and handling of hydrogen also present safety concerns that must be addressed. Hydrogen is highly flammable and explosive, which means that safety systems and protocols must be in place to minimize the risk of accidents.

7. Supply Chain Limitations: The availability of hydrogen is also a challenge. Currently, most of the hydrogen produced is from fossil fuels, which raises questions about the sustainability of the fuel. In addition, the supply chain for hydrogen is still limited, which can hinder the growth of the industry.

Addressing these challenges will require collaboration between public and private stakeholders, as well as continued investment in research and development. With the right policies and incentives in place, the expansion of hydrogen infrastructure can be achieved, paving the way for a more sustainable transportation sector.

The potential of hydrogen production and distribution networks

As the potential for hydrogen fuel cell vehicles grows, so does the need for hydrogen production and distribution infrastructure. The current infrastructure for producing and distributing hydrogen is limited and largely concentrated in a few regions. To support widespread adoption of fuel cell vehicles, significant investment in infrastructure is needed. However, there is potential for hydrogen production and distribution networks to be established that can effectively support this technology.

Hydrogen Production

There are various methods for producing hydrogen, but the most commonly used are steam methane reforming (SMR), which uses natural gas as a feedstock, and electrolysis, which uses renewable energy sources like wind and solar to split water into hydrogen and oxygen. SMR is currently the most commonly used method of hydrogen production, but it is not considered a sustainable solution due to its reliance on natural gas.

The use of electrolysis for hydrogen production has gained momentum in recent years due to the decreasing costs of renewable energy sources. As the cost of renewable energy sources continues to decrease, electrolysis has the potential to become a more viable solution for sustainable hydrogen production. In addition, there are other methods being developed for hydrogen production, such as biological and chemical methods, which may offer more sustainable solutions.

Hydrogen Distribution

The current distribution of hydrogen is limited to a few regions, with the majority of stations located in California, Japan, and Germany. The lack of hydrogen distribution infrastructure is a major barrier to widespread adoption of fuel cell vehicles, as it limits the range of the vehicles and the ability to refuel.

To effectively support fuel cell vehicles, a distribution network must be established that can transport hydrogen from production facilities to refueling stations. The use of pipelines for hydrogen distribution is being explored, but it is currently limited due to safety concerns and the high cost of pipeline construction. The most common method for hydrogen distribution is through trucks that transport compressed hydrogen to refueling stations.

The potential for hydrogen distribution networks is not limited to fuel cell vehicles. Hydrogen can also be used for stationary power generation and in industrial applications, such as steel production and chemical manufacturing. These industries can serve as potential customers for hydrogen production and distribution networks, which can help to drive down costs and increase the viability of the technology.

Challenges

The establishment of hydrogen production and distribution networks faces several challenges. The first challenge is the high cost of infrastructure development, which includes the construction of production facilities, pipelines, and refueling stations. The second challenge is the need for large-scale investment to support the development of these networks. This investment must come from both the private and public sectors.

The third challenge is the need for standardization in hydrogen production and distribution. Standardization can help to ensure that all vehicles and infrastructure are compatible and can help to drive down costs. Standardization efforts are currently underway, but they are still in the early stages.

Another challenge is the perception of hydrogen as a dangerous fuel. Hydrogen has a low ignition energy and can ignite at low concentrations in air, which has raised concerns about its safety. However, proper handling and storage practices can mitigate these risks.

Conclusion

The potential for hydrogen production and distribution networks to support the widespread adoption of fuel cell vehicles is significant. The development of sustainable methods for hydrogen production, along with the establishment of a distribution network, can help to drive down costs and increase the viability of fuel cell technology. However, significant investment and standardization efforts are needed to address the current limitations of hydrogen infrastructure.

The role of government and private sector in infrastructure development

The development of hydrogen infrastructure is critical to the success of hydrogen fuel cell vehicles. However, the high cost of infrastructure development has been one of the main barriers to the adoption of hydrogen fuel cell vehicles. To overcome this challenge, there is a need for collaboration between the government and the private sector.

The Role of Government in Infrastructure Development

Governments can play a crucial role in the development of hydrogen infrastructure. They can provide funding for research and development, tax incentives for the production and sale of hydrogen fuel cell vehicles, and grants for the construction of refueling stations. Government support can also help to attract private investment in the sector.

One example of government support is the California Fuel Cell Partnership (CaFCP). The CaFCP was established in 1999 to promote the commercialization of hydrogen fuel cell vehicles in California. The partnership is made up of automakers, energy companies, and government agencies. It has played a key role in the development of the hydrogen infrastructure in California, which now has over 40 operational hydrogen refueling stations.

In addition to providing funding, governments can also mandate the use of hydrogen fuel cell vehicles in public fleets. This can create a demand for the vehicles and help to jumpstart the development of the necessary infrastructure. For example, in 2019, the California Air Resources Board (CARB) mandated that all transit agencies in California must transition to zero-emission buses by 2040. This has led to an increase in the adoption of hydrogen fuel cell buses and the development of more refueling stations.

The Role of Private Sector in Infrastructure Development

The private sector also has an important role to play in the development of hydrogen infrastructure. Companies can invest in the construction of refueling stations, hydrogen production facilities, and distribution networks. They can also partner with automakers to develop and produce hydrogen fuel cell vehicles.

One example of private sector investment is the partnership between Toyota and Shell. In 2017, the two companies announced a partnership to expand hydrogen refueling infrastructure in California. The partnership aims to develop seven new hydrogen refueling stations in the state.

Another example is the partnership between Hyundai and Air Liquide. In 2020, the two companies announced a partnership to develop hydrogen refueling stations in Europe. The partnership aims to develop 1,600 hydrogen refueling stations across Europe by 2030.

Collaboration between the Government and Private Sector

To accelerate the development of hydrogen infrastructure, collaboration between the government and private sector is crucial. Government support can help to attract private investment in the sector, while private investment can help to drive innovation and accelerate the development of the necessary infrastructure.

One example of collaboration is the H2USA partnership. H2USA is a public-private partnership that was launched in 2013 by the U.S. Department of Energy (DOE) and automakers. The partnership aims to accelerate the development of hydrogen infrastructure in the United States. It has helped to bring together automakers, energy companies, and government agencies to share knowledge and resources.

Another example is the Hydrogen Council, a global initiative launched in 2017. The Hydrogen Council is made up of over 80 leading companies from the energy, transport, and industrial sectors. The council aims to promote the adoption of hydrogen as a key part of the transition to a low-carbon economy. Its members have pledged to invest $10 billion in hydrogen-related projects over the next five years.

Conclusion

The development of hydrogen infrastructure is critical to the success of hydrogen fuel cell vehicles. While the high cost of infrastructure development has been a barrier to adoption, collaboration between the government and private sector can help to overcome this challenge. Government support can help to attract private investment and reduce the financial risk for businesses interested in developing hydrogen infrastructure. This can take the form of financial incentives, tax credits, and grants for infrastructure development. The private sector, in turn, can bring expertise in the design, construction, and operation of infrastructure, as well as access to capital and technology. Collaborative efforts between the government and private sector have already begun to show promise, with the construction of new hydrogen refueling stations and the expansion of existing networks in some regions. However, continued investment and collaboration will be necessary to create a robust and reliable hydrogen infrastructure that can support the widespread adoption of hydrogen fuel cell vehicles.

Chapter 5: The Future of Hydrogen in Transportation

The potential for widespread adoption of hydrogen fuel cell vehicles

The potential for widespread adoption of hydrogen fuel cell vehicles is a topic of much discussion in the transportation industry. While the technology has advanced significantly over the years, the adoption of fuel cell vehicles has been slow due to various reasons, including infrastructure development and cost-effectiveness. However, with recent advancements in hydrogen infrastructure, fuel cell technology, and growing concerns about the environmental impact of fossil fuels, the potential for widespread adoption of hydrogen fuel cell vehicles is becoming more feasible.

One significant advantage of hydrogen fuel cell vehicles is their potential to provide a zero-emissions solution to transportation. Unlike conventional gasoline-powered vehicles that emit harmful pollutants, fuel cell vehicles only emit water and heat. This makes them an attractive alternative for environmentally conscious consumers, governments, and businesses looking to reduce their carbon footprint.

Additionally, hydrogen fuel cell vehicles have a longer driving range compared to battery-electric vehicles. This is because fuel cells generate electricity by combining hydrogen with oxygen to produce water, which generates electricity. The process is continuous, allowing for a constant supply of electricity, making it ideal for long-distance travel. Furthermore, hydrogen fueling stations take less time to refuel compared to battery-electric vehicles, which require lengthy charging times. This makes hydrogen fuel cell vehicles a more practical solution for consumers who require long-distance travel.

Another factor that could contribute to the widespread adoption of hydrogen fuel cell vehicles is the development of renewable sources of hydrogen production. Currently, most of the hydrogen produced is derived from natural gas, which is a fossil fuel. However, there is increasing interest in developing renewable sources of hydrogen, such as wind and solar energy, which would further reduce the environmental impact of hydrogen production and use.

Furthermore, government policies and incentives could also play a significant role in the adoption of hydrogen fuel cell vehicles. Governments around the world are implementing various policies and incentives to encourage the adoption of clean energy vehicles. For example, in California, the state offers rebates of up to $4,500 for the purchase of fuel cell vehicles. Similarly, in Japan, the government has set a goal of having 800,000 fuel cell vehicles on the road by 2030, and is offering tax incentives and subsidies for fuel cell vehicle purchases.

However, there are still significant challenges to overcome before hydrogen fuel cell vehicles can achieve widespread adoption. One of the most significant challenges is the cost of fuel cell vehicles. While the cost of fuel cell technology has reduced significantly over the years, it is still more expensive than conventional gasoline-powered vehicles. Similarly, the cost of hydrogen production and infrastructure development is also high.

Moreover, the lack of hydrogen fueling infrastructure is still a significant barrier to the widespread adoption of hydrogen fuel cell vehicles. As of 2021, there are only a few hundred hydrogen fueling stations globally, with most of them concentrated in California, Japan, and Germany. To encourage widespread adoption, a more extensive network of fueling stations needs to be developed, and this requires significant investment.

In conclusion, the potential for widespread adoption of hydrogen fuel cell vehicles is significant, and there are various factors that could contribute to this adoption. With the development of renewable sources of hydrogen production, government policies and incentives, and advancements in fuel cell technology and infrastructure, the future of hydrogen in transportation is becoming more promising. However, significant challenges remain, and collaboration between governments, private sector, and consumers is essential to overcome these challenges and promote the widespread adoption of hydrogen fuel cell vehicles.

The impact of technological advancements and research on hydrogen fuel cell technology

Technological advancements and research play a vital role in the development of hydrogen fuel cell technology. Scientists and engineers are continuously working to improve the efficiency and performance of fuel cells, as well as reduce their cost.

One area of research is the development of new materials for fuel cells. Currently, most fuel cells use platinum as a catalyst, which is expensive and rare. Researchers are exploring alternative catalysts, such as iron or nickel, that could be more affordable and abundant. Additionally, they are investigating new materials for the membranes and electrodes in fuel cells, which could improve durability and efficiency.

Another area of research is the development of new fuel cell designs. For example, researchers are exploring the use of solid oxide fuel cells, which operate at higher temperatures and have the potential to be more efficient than current fuel cell designs. Solid oxide fuel cells can also be powered by a wider range of fuels, including natural gas and biofuels.

Advancements in hydrogen production and storage technologies are also important for the future of hydrogen in transportation. Research is being done on various methods of producing hydrogen, including electrolysis of water using renewable energy sources such as wind and solar power. New storage technologies are also being developed to address the challenges of storing and transporting hydrogen, such as metal hydrides and complex chemical compounds.

Additionally, advancements in automation and digitalization could improve the efficiency of hydrogen fuel cell vehicles. For example, advanced control systems and sensors could optimize the performance of fuel cells and reduce maintenance requirements. Digital platforms could also enable the integration of hydrogen fuel cell vehicles into smart grids and transportation networks.

The impact of these technological advancements and research on hydrogen fuel cell technology could be significant. They could lead to more efficient, affordable, and durable fuel cells, as well as improvements in hydrogen production and storage technologies. As a result, the widespread adoption of hydrogen fuel cell vehicles could become more feasible, and the environmental benefits of hydrogen as a fuel source could be maximized. However, it is important to continue investing in research and development to ensure that hydrogen fuel cell technology continues to improve and meet the needs of the market.

The role of hydrogen in a broader shift towards clean energy transportation

As the world becomes increasingly aware of the need to reduce greenhouse gas emissions and combat climate change, clean energy transportation has become a top priority for many governments and organizations. Hydrogen fuel cell technology has the potential to play a significant role in this shift towards clean energy transportation.

One of the main advantages of hydrogen fuel cell vehicles is that they emit only water vapor as a byproduct, making them a zero-emission vehicle. As such, hydrogen fuel cell technology is seen as a promising alternative to traditional gasoline and diesel vehicles, which are significant contributors to air pollution and greenhouse gas emissions. In addition, hydrogen fuel cell vehicles have a longer driving range than electric vehicles, making them a more viable option for long-distance travel.

Another advantage of hydrogen fuel cell technology is that it can be used in a variety of transportation modes, including cars, buses, trains, and even ships and airplanes. As such, hydrogen fuel cell technology has the potential to transform a broad range of industries and sectors, not just the automotive industry.

The shift towards clean energy transportation is being driven by a number of factors, including government regulations, consumer demand, and technological advancements. Many countries and cities around the world have set ambitious targets for reducing greenhouse gas emissions and transitioning to clean energy transportation. For example, the European Union has set a target of reducing greenhouse gas emissions from transportation by 90% by 2050, while California has set a target of having 1.5 million zero-emission vehicles on its roads by 2025.

Technological advancements are also playing a key role in the shift towards clean energy transportation. As hydrogen fuel cell technology continues to improve and become more efficient, it is becoming a more viable option for widespread adoption. Researchers and engineers are working to improve the performance and durability of fuel cells, reduce their cost, and improve the efficiency of hydrogen production and storage.

In addition to transportation, hydrogen fuel cell technology is also being explored as a potential source of clean energy for other sectors, such as power generation and industrial processes. This could further increase the demand for hydrogen and accelerate the development of hydrogen infrastructure.

Overall, hydrogen fuel cell technology has the potential to play a significant role in a broader shift towards clean energy transportation. While there are still challenges to be overcome, such as infrastructure development and cost-effectiveness, the increasing awareness of the need to reduce greenhouse gas emissions and combat climate change is driving greater investment and research into hydrogen fuel cell technology. As such, the future looks bright for hydrogen in transportation and beyond.

The challenges and opportunities of a hydrogen-based transportation system

As the world continues to seek out alternative energy sources to reduce carbon emissions and combat climate change, hydrogen is emerging as a promising candidate for powering transportation. However, the widespread adoption of hydrogen fuel cell vehicles comes with both challenges and opportunities.

One of the most significant challenges is the need for infrastructure development to support a hydrogen-based transportation system. This includes the construction of refueling stations, as well as production and distribution networks for hydrogen. The cost of this infrastructure can be high, and without sufficient government and private sector support, it may be difficult to achieve.

Another challenge is the cost-effectiveness of hydrogen fuel cell vehicles compared to traditional gasoline vehicles. While the technology has come a long way in terms of efficiency and cost, it still remains more expensive than gasoline vehicles. However, as the technology continues to evolve and production scales up, the cost is expected to decrease.

Despite these challenges, there are also significant opportunities associated with a hydrogen-based transportation system. One of the most significant benefits is the reduction in carbon emissions. Hydrogen fuel cell vehicles emit only water vapor and heat, making them a zero-emission alternative to traditional gasoline vehicles. This is a significant advantage in a world where reducing carbon emissions is critical to mitigating climate change.

Another opportunity is the potential for increased energy security. Hydrogen can be produced from a variety of sources, including renewable sources such as wind and solar. This means that a hydrogen-based transportation system could reduce our reliance on foreign oil and increase our energy independence.

In addition, hydrogen fuel cell vehicles offer a high level of performance and efficiency. They have a longer range than electric vehicles and can be refueled in minutes, making them a more practical option for long-distance travel. They are also highly efficient, converting up to 60% of the energy from hydrogen into usable power, compared to the 20% efficiency of gasoline vehicles.

Finally, a hydrogen-based transportation system could offer new opportunities for economic growth and job creation. The development of infrastructure and technology would require significant investment, and the growth of the hydrogen industry could create new jobs in manufacturing, research and development, and maintenance.

In conclusion, the challenges associated with a hydrogen-based transportation system should not be underestimated, but neither should the opportunities. A hydrogen-based transportation system has the potential to significantly reduce carbon emissions, increase energy security, and create new economic opportunities. As technology continues to evolve and infrastructure develops, it is possible that we will see widespread adoption of hydrogen fuel cell vehicles in the coming years.

Chapter 6: Case Studies in Hydrogen Power
Real-world examples of hydrogen fuel cell vehicles and infrastructure

Hydrogen fuel cell technology has been implemented in a variety of real-world settings, ranging from personal vehicles to public transportation and even maritime vessels. These case studies demonstrate the feasibility and potential benefits of a hydrogen-based transportation system.

One notable example is the Toyota Mirai, a hydrogen fuel cell vehicle that has been commercially available since 2014. The Mirai has a range of approximately 312 miles and refueling time of around 5 minutes, making it a viable alternative to gasoline-powered vehicles. Toyota has invested heavily in hydrogen infrastructure, collaborating with partners to build refueling stations in California and other locations.

Another example of hydrogen fuel cell vehicles is the fleet of buses operated by the Orange County Transportation Authority in California. The agency began testing fuel cell buses in 2006 and now operates a fleet of 24 buses powered by hydrogen fuel cells. These buses emit only water vapor and have a range of approximately 300 miles on a single tank of hydrogen. The success of this program has led to the expansion of fuel cell bus fleets in other cities around the world, including London and Tokyo.

Hydrogen fuel cell technology has also been implemented in maritime vessels. For example, the MS Viking Grace, a passenger ferry that operates in the Baltic Sea, is powered by a combination of hydrogen fuel cells and batteries. This system provides clean, efficient power for the vessel, reducing emissions and operating costs. The success of the Viking Grace has inspired other maritime companies to explore the potential of hydrogen fuel cell technology for their own vessels.

In addition to vehicles, hydrogen fuel cell technology has also been used to power buildings and provide backup power during emergencies. For example, the University of California, Irvine has a microgrid that incorporates hydrogen fuel cells, solar panels, and batteries to provide clean, reliable power to its campus. The microgrid can operate independently from the grid during power outages, providing a critical source of energy during emergencies.

Overall, these case studies demonstrate the potential of hydrogen fuel cell technology to power a variety of vehicles and applications. As infrastructure continues to develop and costs decrease, it is likely that we will see more widespread adoption of this technology in the future.

The success stories and challenges of hydrogen-powered transportation

Hydrogen-powered transportation has seen success in various parts of the world, but it has also encountered challenges. Here are some examples of success stories and challenges in the development of hydrogen-powered transportation.

One success story is Japan's Mirai project, which aims to promote the use of hydrogen fuel cell vehicles as part of a sustainable society. The project involves the installation of hydrogen stations across the country, the development of fuel cell vehicles, and the promotion of hydrogen-based industries. The project has been successful in promoting hydrogen as a clean energy source and in increasing the number of hydrogen fuel cell vehicles on the road.

Another success story is California's Zero-Emission Vehicle (ZEV) program, which aims to reduce greenhouse gas emissions by promoting the use of zero-emission vehicles, including hydrogen fuel cell vehicles. The program has incentivized the development of hydrogen refueling infrastructure, resulting in the installation of over 40 hydrogen stations across the state. California also has the highest number of hydrogen fuel cell vehicles on the road in the United States.

Germany's H2 Mobility initiative is another example of a successful hydrogen infrastructure development project. The initiative aims to deploy 100 hydrogen stations across Germany by 2023, with the goal of having 1 million fuel cell vehicles on the road by 2030. The project has already installed over 80 hydrogen stations, making Germany the country with the second-highest number of hydrogen refueling stations in Europe.

However, there have also been challenges in the development of hydrogen-powered transportation. One major challenge is the high cost of hydrogen infrastructure development, which can be a barrier to widespread adoption. In addition, the limited range of some fuel cell vehicles can be a challenge for long-distance travel, although newer models have addressed this issue.

Another challenge is the production and transportation of hydrogen. The majority of hydrogen produced today is from fossil fuels, which raises concerns about its carbon footprint. However, there are alternative methods of hydrogen production, such as electrolysis using renewable energy sources, which could reduce the carbon footprint of hydrogen.

Safety concerns are also a challenge in the development of hydrogen-powered transportation.

Hydrogen is a highly flammable gas, and there have been concerns about the safety of hydrogen refueling stations and fuel cell vehicles. However, rigorous safety standards and protocols have been put in place to ensure the safe operation of hydrogen infrastructure and vehicles.

Overall, the development of hydrogen-powered transportation has seen both successes and challenges. While there is still work to be done in the expansion of infrastructure and the reduction of costs, hydrogen has the potential to play a significant role in the transition to a cleaner and more sustainable transportation system.

The potential for future applications in various industries

Hydrogen fuel cell technology is not limited to transportation applications. Its potential extends to a variety of other industries, including energy generation, manufacturing, and more. As such, it is important to examine the potential for future applications of hydrogen in these industries.

One major area of potential for hydrogen is in energy generation. Hydrogen fuel cells can be used to generate electricity with high efficiency and low emissions, making it an attractive alternative to traditional energy sources such as coal and natural gas. One potential application is in the development of hydrogen-powered homes, where fuel cells can provide electricity and heat without producing harmful emissions.

Another industry where hydrogen has potential is in manufacturing. Hydrogen can be used in various industrial processes, including the production of chemicals, metals, and other materials. In fact, hydrogen is already used in some manufacturing processes, such as in the production of ammonia for fertilizers. Using hydrogen instead of traditional energy sources in these processes can reduce emissions and improve efficiency.

Hydrogen can also be used in the transportation of goods, such as in the shipping industry. Hydrogen fuel cell technology can be used to power ships and reduce emissions, helping to reduce the industry's impact on the environment. This potential has already been recognized by some shipping companies, with several working to develop hydrogen-powered ships.

Another area where hydrogen has potential is in the aviation industry. While battery-powered electric planes have been proposed, they are not yet viable for large commercial aircraft. Hydrogen fuel cell technology could provide an alternative solution, offering a low-emission power source for airplanes. Several companies, including Airbus and ZeroAvia, are already working on developing hydrogen-powered planes.

There are also potential applications for hydrogen in the space industry. NASA has been exploring the use of hydrogen fuel cells for space exploration since the 1960s, and has used them to power spacecraft and space stations. With the increasing interest in space exploration and tourism, there may be opportunities for further development of hydrogen fuel cell technology in this industry.

While there are clear opportunities for hydrogen fuel cell technology in various industries, there are also challenges that must be addressed. One of the biggest challenges is the cost of hydrogen production and distribution. While there are already existing hydrogen production methods, such as electrolysis and steam methane reforming, they are currently more expensive than traditional energy sources. Additionally, there are still limitations to the current hydrogen distribution infrastructure, which can limit the widespread adoption of hydrogen as an energy source.

Another challenge is the safety concerns associated with hydrogen. Hydrogen is a highly flammable gas, and there have been concerns about the safety of hydrogen fuel cells in transportation applications. However, studies have shown that hydrogen fuel cells are safe when properly designed and maintained.

In conclusion, the potential for hydrogen extends beyond transportation, and there are clear opportunities for its use in various industries. However, there are also challenges that must be addressed, including the cost of production and distribution and safety concerns. Despite these challenges, the increasing interest in clean energy and the potential benefits of hydrogen fuel cell technology make it an area of great promise for future applications.

Chapter 7: Overcoming the Challenges of Hydrogen Power

The barriers to widespread adoption of hydrogen power

Hydrogen power has the potential to revolutionize the energy industry, but there are several barriers to its widespread adoption. These barriers can be broadly classified into technological, economic, and social factors.

One of the major technological barriers is the high cost of fuel cell technology. While the cost of fuel cells has been decreasing steadily over the years, it still remains prohibitively expensive for widespread adoption. This is due in part to the high cost of materials such as platinum, which is used in the catalysts in fuel cells. Additionally, there are issues with the durability and reliability of fuel cells, which can limit their usefulness in certain applications.

Another technological barrier is the lack of infrastructure for producing, transporting, and storing hydrogen. Hydrogen is not readily available and must be produced through a process such as electrolysis or steam methane reforming. The transportation and storage of hydrogen also requires specialized infrastructure, which can be costly to build and maintain.

Economic barriers include the lack of government incentives and subsidies for hydrogen power. Many countries provide incentives for electric vehicles, but the same level of support has not been extended to hydrogen fuel cell vehicles. Additionally, the lack of economies of scale for fuel cell technology means that the cost of producing and using hydrogen is still higher than traditional fossil fuels.

Social barriers include the lack of public awareness and understanding of hydrogen power. Many people are unfamiliar with the technology and may be hesitant to adopt it due to concerns about safety and reliability. Additionally, the existing infrastructure for traditional fossil fuels is already well-established, making it difficult to transition to a new energy source.

To overcome these barriers, a concerted effort is needed from governments, industry, and the public. Government support in the form of subsidies and incentives can help to stimulate the growth of the hydrogen industry. Industry can work to improve the technology and reduce the costs of fuel cells and hydrogen production. Public education campaigns can help to raise awareness and understanding of the benefits of hydrogen power.

Overall, while there are certainly barriers to widespread adoption of hydrogen power, these challenges

can be overcome with a collaborative effort and continued
investment in research and development.

The solutions to address these challenges

Hydrogen power faces several challenges that hinder its widespread adoption as a clean and sustainable source of energy. These include the high cost of producing, storing, and transporting hydrogen, the lack of infrastructure, and the limited availability of hydrogen fuel cell vehicles. However, various solutions have been proposed to address these challenges and accelerate the adoption of hydrogen power.

One of the most promising solutions is the development of new and innovative technologies that can reduce the cost of producing, storing, and transporting hydrogen. For example, advancements in renewable energy sources such as wind and solar power can be used to generate hydrogen through electrolysis, which is a cost-effective and sustainable method of producing hydrogen. Additionally, new materials and designs for hydrogen storage tanks and pipelines can also help to reduce the cost and increase the efficiency of hydrogen storage and transportation.

Another solution is the expansion of infrastructure, particularly the establishment of a network of hydrogen refueling stations. Governments and private investors can work together to provide financial incentives for the development of refueling stations and to support the expansion of hydrogen infrastructure. This can be achieved through grants, tax incentives, and other forms of financial support.

Moreover, collaborations among various stakeholders in the hydrogen value chain are essential to ensure a smooth transition to a hydrogen-based economy. This includes collaborations between governments, research institutions, manufacturers, and investors. Joint ventures and partnerships among companies can also help to accelerate the development and commercialization of hydrogen technologies.

In addition, government policies can play a critical role in accelerating the adoption of hydrogen power. Policy frameworks that provide long-term stability and certainty, such as renewable portfolio standards, tax incentives, and feed-in tariffs, can encourage investments in hydrogen technologies. Governments can also mandate the adoption of hydrogen fuel cell vehicles in certain sectors such as public transportation, which can help to create a market for hydrogen and stimulate demand.

Finally, education and public awareness campaigns are necessary to inform the public about the benefits of hydrogen power and dispel misconceptions about the technology. This can help to increase public support and create a demand for hydrogen-powered vehicles and infrastructure.

In conclusion, the challenges facing the widespread adoption of hydrogen power can be addressed through the development of new and innovative technologies, the expansion of infrastructure, collaborations among stakeholders, government policies, and education and public awareness campaigns. These solutions can help to accelerate the transition to a hydrogen-based economy and achieve a sustainable and clean energy future.

The role of individuals, corporations, and governments in advancing hydrogen power technology and infrastructure

The role of individuals, corporations, and governments in advancing hydrogen power technology and infrastructure is critical for overcoming the challenges associated with widespread adoption of hydrogen power.

Individuals have an important role to play in supporting the development and adoption of hydrogen power technology. This can involve advocating for policies and incentives that promote the use of hydrogen fuel cell vehicles and infrastructure, as well as supporting the development of hydrogen power through their purchasing decisions. Consumers can also provide feedback to automakers and government entities on the effectiveness and convenience of hydrogen fuel cell vehicles and infrastructure, which can help to guide future development.

Corporations also have a significant role in advancing hydrogen power technology and infrastructure. They can invest in research and development of new technologies and materials, as well as in the production and distribution of hydrogen fuel. By collaborating with government agencies and other stakeholders, corporations can help to establish standards and regulations that promote the safe and efficient use of hydrogen power.

Governments are critical to advancing hydrogen power technology and infrastructure, as they can provide funding and support for research and development, as well as for the construction of infrastructure. Governments can also establish policies and regulations that promote the use of hydrogen power, such as tax incentives and subsidies for hydrogen fuel cell vehicles and infrastructure. Additionally, governments can work to establish international agreements and collaborations to promote the development and adoption of hydrogen power technology and infrastructure on a global scale.

In order to advance hydrogen power technology and infrastructure, it is important for individuals, corporations, and governments to work together in a collaborative and coordinated manner. This can involve establishing partnerships and collaborations between different stakeholders, as well as sharing knowledge and resources to advance the development of hydrogen power.

One example of collaboration between governments, corporations, and individuals is the H2USA program, which was launched by the US Department of Energy in 2013 to accelerate the development and adoption of hydrogen fuel cell vehicles and infrastructure. The program brought together stakeholders from the government, industry, and non-profit organizations to collaborate on the development and deployment of hydrogen fuel cell vehicles and infrastructure. Similarly, the European Commission has established the Fuel Cells and Hydrogen Joint Undertaking (FCH JU), which aims to accelerate the commercialization of hydrogen and fuel cell technologies in Europe through public-private partnerships.

In conclusion, the advancement of hydrogen power technology and infrastructure requires the collective efforts of individuals, corporations, and governments. By working together in a collaborative and coordinated manner, we can overcome the challenges associated with widespread adoption of hydrogen power and pave the way for a more sustainable and cleaner future.

Conclusion
The potential of hydrogen as a clean energy solution in transportation

The transportation sector is one of the largest contributors to greenhouse gas emissions, and finding clean energy solutions is critical in the fight against climate change. Hydrogen fuel cell technology is a promising option, offering a zero-emission alternative to fossil fuel-powered vehicles. This book has explored the potential of hydrogen in transportation and its various applications, from cars and trucks to buses and trains. While there are challenges to overcome, such as infrastructure development and cost-effectiveness, the future of hydrogen power looks promising.

One of the main advantages of hydrogen fuel cell vehicles is that they produce zero emissions, meaning they have the potential to significantly reduce greenhouse gas emissions in the transportation sector. As the world continues to move towards decarbonization, hydrogen has the potential to play a key role in achieving net-zero emissions goals. Moreover, hydrogen fuel cell vehicles can offer a range and refueling time comparable to conventional vehicles, making them a viable option for consumers.

Another advantage of hydrogen fuel cell technology is its versatility. Hydrogen can be produced from a variety of sources, including renewable energy such as wind and solar power. This means that it has the potential to be a truly sustainable energy solution, providing a pathway to decarbonization across multiple sectors. In addition to transportation, hydrogen can be used in power generation and industrial processes, further expanding its potential impact.

Despite the advantages of hydrogen fuel cell technology, there are challenges to overcome. One of the main challenges is the cost of infrastructure development and maintenance, including the construction and operation of refueling stations. However, collaboration between the government and private sector can help to overcome this challenge, with government support helping to attract private investment.

Another challenge is the need for technological advancements and research to improve the efficiency and cost-effectiveness of hydrogen fuel cell vehicles. Advancements in fuel cell technology and the development of new materials have the potential to make hydrogen more competitive with traditional fossil fuels, further driving its adoption in transportation and other industries.

Overall, hydrogen fuel cell technology has the potential to play a significant role in the transition to a low-carbon economy. While there are challenges to overcome, such as infrastructure development and cost-effectiveness, the potential benefits of hydrogen make it a promising option for the future. With continued research and investment, the potential of hydrogen as a clean energy solution in transportation and other sectors is only set to grow.

The challenges and opportunities of hydrogen fuel cell technology

Hydrogen fuel cell technology has the potential to revolutionize the transportation industry by providing a clean, efficient, and reliable source of energy for vehicles. However, there are several challenges that must be addressed before widespread adoption can occur.

One of the primary challenges is the cost of developing and implementing the necessary infrastructure, including production, storage, and distribution of hydrogen fuel. The high cost of infrastructure development has been a significant barrier to adoption, but collaboration between the government and private sector can help to overcome this challenge. Government support can help to attract private investment, and partnerships between governments and private companies can help to share the costs and risks of infrastructure development.

Another challenge is the limited availability of refueling stations, which can make it difficult for consumers to use hydrogen fuel cell vehicles for long trips or daily commuting. However, efforts are underway to expand the number of refueling stations, and advancements in technology may also help to increase the range of hydrogen fuel cell vehicles.

There are also concerns about the safety of hydrogen fuel cell vehicles, particularly in the event of an accident. However, studies have shown that hydrogen fuel cell vehicles are no more dangerous than traditional gasoline-powered vehicles, and safety measures are being implemented to address any potential risks.

Despite these challenges, there are significant opportunities for hydrogen fuel cell technology in the transportation industry. Hydrogen fuel cell vehicles offer several advantages over traditional gasoline-powered vehicles, including zero emissions, greater energy efficiency, and reduced dependence on foreign oil.

Furthermore, hydrogen fuel cell technology has the potential to play a key role in the broader shift towards clean energy transportation. As the world seeks to reduce carbon emissions and combat climate change, hydrogen fuel cell technology can help to achieve these goals by providing a clean and sustainable source of energy for vehicles.

In conclusion, while there are challenges that must be overcome, hydrogen fuel cell technology has the potential to transform the transportation industry by providing a clean and sustainable source of energy for vehicles. The development of hydrogen infrastructure, expansion of refueling stations, and advancements in technology are all critical steps towards realizing this potential. With collaboration between governments, private companies, and individuals, hydrogen fuel cell technology can help to create a cleaner, more sustainable future for transportation.

The importance of continued research and development in clean energy transportation

Clean energy transportation is essential for mitigating the impact of climate change, reducing pollution and dependence on fossil fuels, and ensuring a sustainable future. While there are several clean energy options available, hydrogen fuel cell technology has gained significant attention in recent years. The potential of hydrogen as a clean energy solution in transportation is vast, but it also presents several challenges and opportunities that need to be addressed.

One of the main challenges of hydrogen fuel cell technology is the high cost of infrastructure development, including production, storage, and distribution. However, as discussed in the previous chapters, significant progress has been made in advancing hydrogen infrastructure, and further collaboration between the government, private sector, and research institutions can help overcome these challenges. Additionally, technological advancements and economies of scale can significantly reduce the cost of hydrogen infrastructure in the future.

Another challenge of hydrogen fuel cell technology is the lack of public awareness and education about the benefits and potential of hydrogen as a clean energy source.

It is crucial to educate the public and policymakers about the benefits of hydrogen fuel cell technology to encourage its widespread adoption. Further, public-private partnerships can help increase public awareness and promote the use of hydrogen fuel cell technology in transportation.

One of the significant opportunities presented by hydrogen fuel cell technology is its potential to integrate with other clean energy technologies. For example, hydrogen can be produced using renewable energy sources such as wind and solar power. Hydrogen can also be used to store excess renewable energy, which can be used during peak demand periods. Integrating hydrogen with other clean energy technologies can increase efficiency and reduce the overall cost of clean energy production and transportation.

Continued research and development in hydrogen fuel cell technology is essential to further its potential in clean energy transportation. This includes advancements in fuel cell technology to improve efficiency and reduce costs, increasing the durability and lifespan of fuel cells, and improving the production and distribution of hydrogen. The development of innovative materials, such as advanced catalysts, can also help reduce the cost and improve the performance of fuel cells.

In conclusion, hydrogen fuel cell technology presents significant opportunities for clean energy transportation. While there are challenges to overcome, including infrastructure development and public awareness, collaboration between the government, private sector, and research institutions can help address these challenges. Further, the integration of hydrogen with other clean energy technologies and continued research and development can significantly increase the potential of hydrogen as a clean energy solution in transportation.

THE END

Key Terms and Definitions

To help you better understand the language and concepts related to aging and older adults, below you will find a list of key terms and their definitions.

1. Hydrogen fuel cell: a device that converts hydrogen and oxygen into water, producing electricity and heat in the process.

2. Hydrogen infrastructure: the network of hydrogen production, transportation, storage, and dispensing facilities necessary to support the use of hydrogen fuel cell vehicles.

3. Fuel cell vehicle: a vehicle that uses a hydrogen fuel cell to power an electric motor, producing only water as a byproduct.

4. Electrolysis: a process in which an electric current is used to split water molecules into hydrogen and oxygen gas.

5. Renewable hydrogen: hydrogen produced from renewable sources such as wind, solar, or hydro power.

6. Carbon capture and utilization (CCU): a process that captures carbon dioxide emissions and converts them into valuable products such as synthetic fuels or chemicals.

7. Green hydrogen: hydrogen produced from renewable sources using electrolysis powered by renewable energy.

8. Blue hydrogen: hydrogen produced from fossil fuels, but with carbon capture and utilization to reduce emissions.

9. Grey hydrogen: hydrogen produced from fossil fuels without carbon capture and utilization, resulting in high greenhouse gas emissions.

10. Hydrogen storage: the process of storing hydrogen gas in a safe and efficient manner, usually through compression, liquefaction, or absorption into materials such as metal hydrides.

Supporting Materials

Introduction

- International Energy Agency. (2020). The future of hydrogen: Seizing today's opportunities. https://www.iea.org/reports/the-future-of-hydrogen

- U.S. Department of Energy. (2021). Hydrogen and fuel cells. https://www.energy.gov/eere/fuelcells/hydrogen-and-fuel-cells

Chapter 1

- Fuel Cells and Hydrogen Joint Undertaking. (2019). The fuel cell and hydrogen observatory. https://www.fchobservatory.eu/

- National Renewable Energy Laboratory. (2019). Fuel cell technology status and future prospects. https://www.nrel.gov/docs/fy19osti/72487.pdf

Chapter 2

- National Academies of Sciences, Engineering, and Medicine. (2018). Review of the research program of the U.S. DRIVE Partnership: Fifth report. The National Academies Press. https://doi.org/10.17226/25196

- U.S. Department of Energy. (2017). A retrospective of the U.S. Department of Energy's flagship research programs in hydrogen production, storage, and fuel cells: A report to Congress.

https://www.energy.gov/sites/prod/files/2017/07/f35/H2_
Retrospective_Report_Final_7-19-17.pdf

Chapter 3

- European Alternative Fuels Observatory. (2021). Hydrogen
fuel cell vehicles. https://www.eafo.eu/knowledge-
base/alternative-fuels/hydrogen

- U.S. Department of Energy. (2021). Benefits and
considerations of electricity as a vehicle fuel.
https://www.energy.gov/eere/electricvehicles/benefits-and-
considerations-electricity-vehicle-fuel

Chapter 4

- National Renewable Energy Laboratory. (2020). H2@Scale
analysis: 2020 update.
https://www.nrel.gov/docs/fy20osti/75165.pdf

- U.S. Department of Energy. (2019). Hydrogen and fuel cell
technologies: Infrastructure considerations of hydrogen as a
transportation fuel.
https://www.energy.gov/eere/fuelcells/hydrogen-and-fuel-
cell-technologies-infrastructure-considerations-hydrogen

Chapter 5

- European Commission. (2020). European hydrogen
strategy: A hydrogen-powered economy. https://eur-
lex.europa.eu/legal-

content/EN/TXT/PDF/?uri=CELEX:52020DC0301&from=
EN

- Toyota Motor Corporation. (2021). Toyota's vision for a
hydrogen society.
https://global.toyota/en/story/23958944.html
Chapter 6
- Fuel Cells and Hydrogen Joint Undertaking. (2019).
Hydrogen mobility Europe: Commercialisation study of
hydrogen powered fuel cell trucks.
https://www.fch.europa.eu/sites/default/files/H2ME_Truc
k_Study_Summary_June2019.pdf
- Hydrogen Council. (2021). Hydrogen in ports and
maritime. https://hydrogencouncil.com/en/hydrogen-in-
ports-and-maritime/
Chapter 7
- California Energy Commission. (2019). California hydrogen
infrastructure outlook.
https://ww2.arb.ca.gov/sites/default/files/2021-
02/2019_hydrogen_infrastructure_outlook.pdf
- National Renewable Energy Laboratory. (2021). Hydrogen
station cost projections.
https://www.nrel.gov/docs/fy21osti/79713.pdf
Conclusion

- European Commission. (2021). Hydrogen strategy for a climate-neutral Europe. https://eur-lex.europa.eu/legal-content/EN/TXT/PDF/?uri=CELEX